MW01124115

The Weather Watcher

by David Conrad

Content and Reading Adviser: Joan Stewart
Educational Consultant/Literacy Specialist
New York Public Schools

Spyglass
BOOKS

COMPASS POINT BOOKS

Minneapolis, Minnesota

Compass Point Books
3722 West 50th Street, #115
Minneapolis, MN 55410

Visit Compass Point Books on the Internet at *www.compasspointbooks.com*
or e-mail your request to *custserv@compasspointbooks.com*

Photographs ©:
Two Coyote Studios/Mary Walker Foley, cover, 4; DigitalVision, 5; Visuals Unlimited/Tom Uhlman, 7;
PhotoDisc, 9, 11 (sunset); Two Coyote Studios/Mary Walker Foley, 11 (pinecone); PhotoDisc, 12; Two
Coyote Studios/Mary Walker Foley, 13, 15, 16, 17, 19.

Project Manager: Rebecca Weber McEwen
Editor: Jennifer Waters
Photo Researcher: Jennifer Waters
Photo Selectors: Rebecca Weber McEwen and Jennifer Waters
Designer: Mary Walker Foley

Library of Congress Cataloging-in-Publication Data

Conrad, David.
  The weather watcher / by David Conrad ; editors, Rebbecca Weber
McEwen, Alison Auch, and Jennifer Waters.
      p. cm. -- (Spyglass books)
Includes bibliographical references and index.
  ISBN 0-7565-0247-0
  1. Meteorology--Juvenile literature. [1. Weather. 2. Meteorology.] I.
McEwen, Rebecca. II. Auch, Alison. III. Waters, Jennifer. IV. Title. V.
Series.
  QC863.5 .C66 2002
  551.5--dc21
                              2001007319

© 2002 by Compass Point Books
All rights reserved. No part of this book may be reproduced without written permission from the publisher.
The publisher takes no responsibility for the use of any of the materials or methods described in this book,
nor for the products thereof.
Printed in the United States of America.

Contents

The Best Show in Town

Nature puts on a big show almost every day—whether the weather is hot and sunny or cold and snowy.

People like to know what weather to expect each day.

Scientists collect information about temperature, wind, *humidity*, and clouds.

They put this information into computers that show what weather might happen under those conditions.

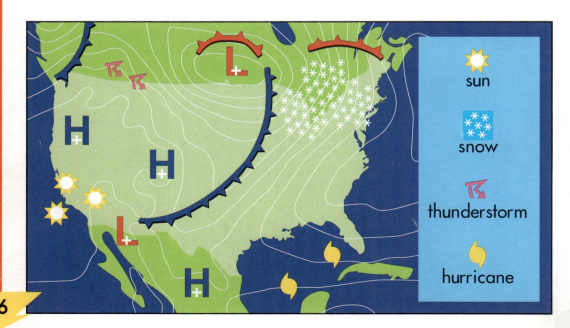

sun

snow

thunderstorm

hurricane

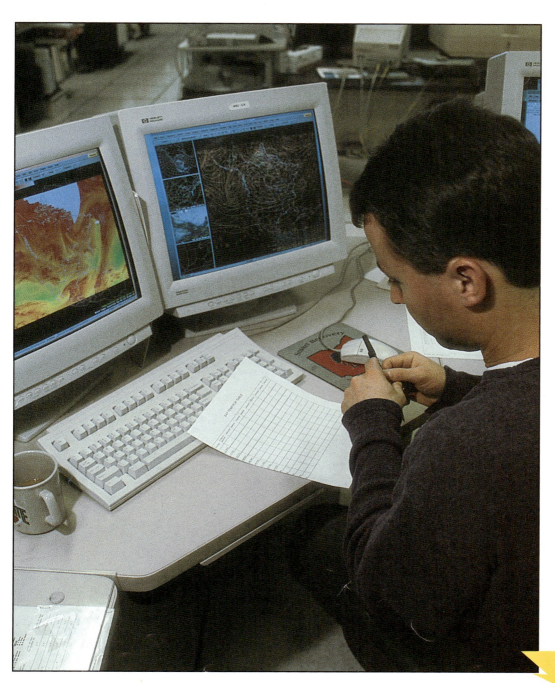

Where in the World?

Different areas of the world have different weather.

Close to the middle of the planet, it is always hot. Close to the *poles*, it is always cold. Areas in between have warm summers and cool winters.

North Pole

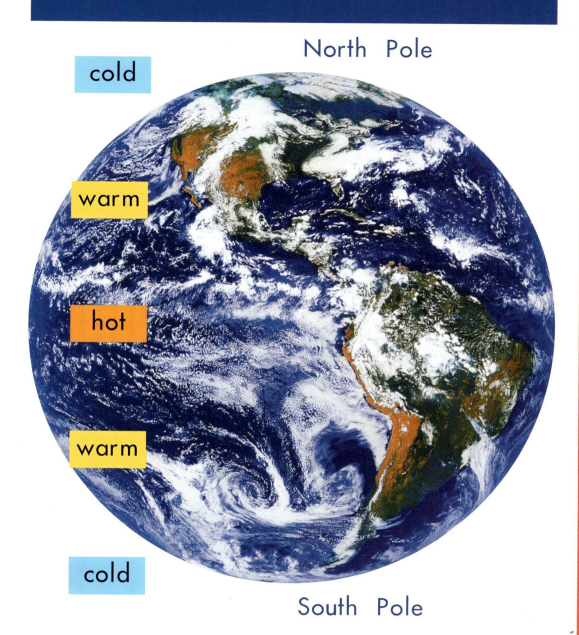

cold

warm

hot

warm

cold

South Pole

9

Nature's Forecasters

Long ago people watched nature for clues about the weather.

The saying "Red sky in morning, sailor take warning" means it will rain.
If the sun looks red, there is dust and water in the air that will fall to the ground as rain.

Did You Know?

If the scales of a pinecone are open, the weather will be dry.

People can *predict* weather
by watching animals.

Deer and elk come down from
the high country before
a big snowstorm.
Cows and geese call to
each other before a storm.

Did You Know?

Grasshoppers chirp more during
warm, dry weather.

Measure the Weather

As you watch the weather, you will start to see *patterns*. You can even make your own weather station at home.

First set up a *thermometer* out of the direct sun. Add other simple tools such as a *rain gauge* or a *wind sock*.

Look and Listen

The best tools of all are your eyes and ears.

If you see clouds building in the west, a storm might be coming. A halo around the sun or moon could mean a storm is coming, too.

Make a Rain Gauge

1. Set a cup out in the open.

2. After a rain, put a ruler in the cup.

3. Measure how much rain fell.

4. Record your measurements and the date on a chart.

5. Do this for one month.

6. Talk about what you noticed.

Make a Weather Chart

Monday	72°	
Tuesday	60°	
Wednesday	48°	
Thursday	32°	
Friday	65°	

Over one week, write the name of the day, the day's high temperature, and draw a picture of what the weather was like.

A Joke!

Some people say a rock is
the best weather tool of all:

If it's dry, the weather's clear.
If it's wet, it's raining.
If it's white, it's snowing.
If it's gone, there's a tornado!

Glossary

humidity—the amount of water in the air

patterns—things that happen the same way

poles—the very top and bottom of the planet

predict—to guess what will happen

rain gauge—something that measures how much rain has fallen

thermometer—a tool that measures how hot or cold something is

wind sock—a bag that wind fills up when it is blowing. The sock points in the direction the wind is blowing.

Learn More

Books

Dussling, Jennifer. *Pink Snow and Other Weird Weather.* Illustrated by Heidi Petach. New York: Grosset & Dunlap, 1998.

Fowler, Allan. *What's the Weather Today?* Chicago: Childrens Press, 1991.

Wallace, Karen. *Whatever the Weather.* New York: DK Publishing, 1999.

Web Sites

Brain Pop
www.brainpop.com/science/seeall.weml (click on "hurricanes")

Kidstorm
www.skydiary.com/kids/

Index

GR: I
Word Count: 245

From David Conrad

I am a scientist who lives in Colorado. I like to climb mountains, square dance, and play with my pet frog, Clyde.